Air Power

AH-1Z VIPER

MEGAN COOLEY PETERSON

BLACK RABBIT BOOKS

Bolt is published by Black Rabbit Books
P.O. Box 3263, Mankato, Minnesota, 56002.
www.blackrabbitbooks.com

Jennifer Besel & Marysa Storm, editors;
Grant Gould, designer; Omay Ayres,
photo researcher

Library of Congress Cataloging-in-Publication Data
Names: Peterson, Megan Cooley, author.
Title: AH-1Z Viper / by Megan Cooley Peterson.
Description: Mankato, Minnesota : Black Rabbit Books, [2020] | Series: Bolt. Air power | Includes bibliographical references and index. | Audience: Age: 8-12. | Audience: Grade: 4 to 6.
Identifiers: LCCN 2018030445 (print) | LCCN 2018038221 (ebook) | ISBN 9781680727920 (e-book) | ISBN 9781680727869 (library binding) | ISBN 9781644660096 (paperback)
Subjects: LCSH: HueyCobra (Helicopter)—Juvenile literature.
Classification: LCC UG1233 (ebook) | LCC UG1233 .P48 2020 (print) | DDC 623.74/63—dc23
LC record available at https://lccn.loc.gov/2018030445

Printed in the United States. 1/19

Image Credits

29palms.marines.mil: Pfc. William Chockey/Marine Corps Air Ground Combat Center, 26 (t); Alamy: APFootage, 10, 26–27; army.mil: Christian Marquardt, 7th Army JMTC Public Affairs/US Army, 23; bellflight.com: Bell, 22; bellflight.com/military: Bell Helicopters Textron Inc., Cover, 15; cgtrader.com, 3Dhorse, 20–21; commons.wikimedia.org: David Monniaux, 17 (l); Jetijones, 12–13 (bkgd), 18–19; Looper5920, 12 (b); Stahlkocher, 16 (b); commons.wikimedia.org/marines.mil: Cpl. Preston Reed/US Marine Corps, 4–5; commons.wikimedia.org/navy.mil: Photographer's Mate 2nd Class Daniel J. McLain/US Navy, 31; flickr.com: Abdullah Nav, 24–25; helis.com/marines.mil: US Marine Corps, 1, 28; imgur.com: nihilist, 3, 18; marines.mil: Cpl. Harley Robinson/US Marine Corps, 27 (b); JMA/US Marines, 32; Lance Cpl. Andy Martinez/US Marine Corps, 13; Sgt Alex Kouns/US Marine Corps, 12 (t); US Marine Corps, 16 (t); mod.gov.cn: Sun Liwei, 9; navy.mil, Mass Communication Specialist 3rd Class Jesse Monford/US Navy, 6; ncsist.org, NCSIST, 17 (r); Shutterstock: BNK Maritime Photographer, 11; inspired-fiona, 19 (bkgd); Melkor3D, 16 (c)

CONTENTS

CHAPTER 1

The Viper's First FLIGHT

It was a cool morning in December 2000. An AH-1Z Viper attack helicopter sat on the runway. Then its engines roared. The chopper lifted off the ground.

FLY 1

A Success

The Viper took off into the air. It zoomed toward the ground. Just before it hit the ground, the chopper pulled up. It **hovered** above the runway. The Viper's first test flight was a success.

The chopper's first flight lasted 15 minutes.

CHAPTER 2

HISTORY of the AH-1Z Viper

In the 1960s, the U.S. military built the Cobra attack helicopter. Soldiers used it during the **Vietnam War**. U.S. Marines began using the Cobra in 1969. They added improvements to the chopper over the years. But by the 1990s, the chopper needed big updates.

UNITED
SHEPHERD

An Updated Chopper

The Marines needed an updated attack helicopter. It had to fly farther than the Cobra. It also had to carry more weapons. **Engineers** began making big changes to the Cobra. The updated chopper became the AH-1Z Viper.

AH-1Z Viper	vs.	AH-1W Super Cobra
2	crew	2
18,500 pounds (8,391 kilograms)	maximum weight	14,750 pounds (6,690 kg)
151 miles (243 kilometers)	distance it can go before refueling	67 miles (108 km)

VIPER TIMELINE

February
2006
Viper completes flight testing.

1995

December
2000
Viper completes first test flight.

November
2016
Vipers begin arriving at the Marines' base in Japan. They replace old Cobras.

2025

September
2010
Viper is ready for **combat**.

2020
The Marines plan to have 189 total Vipers.

CHAPTER 3

Viper FEATURES

The Viper has four **rotors**. The Cobra only has two. The extra rotors help the Viper fly faster. It can fly up to 230 miles (370 km) per hour. The rotors also allow the Viper to carry more weapons.

The AH-1Z is nicknamed "Zulu."

Some of the Viper's Weapons

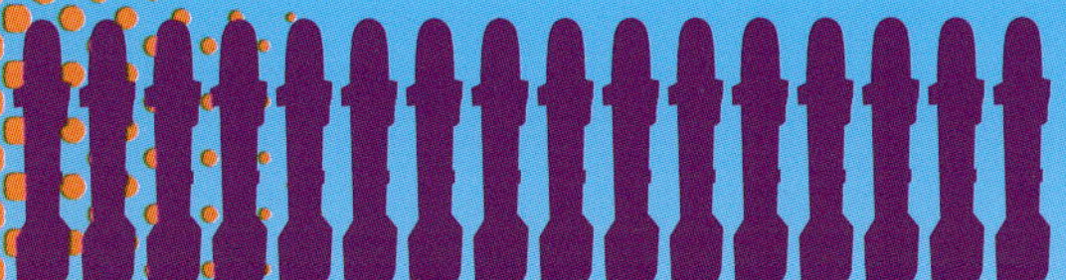

up to 16 Hellfire air-to-ground missiles

Weapons

The Viper is a deadly aircraft. The chopper's cannon can shoot 750 rounds before reloading. The chopper can also fire **missiles** at other aircraft. Shooting down enemy aircraft keeps ground soldiers safer.

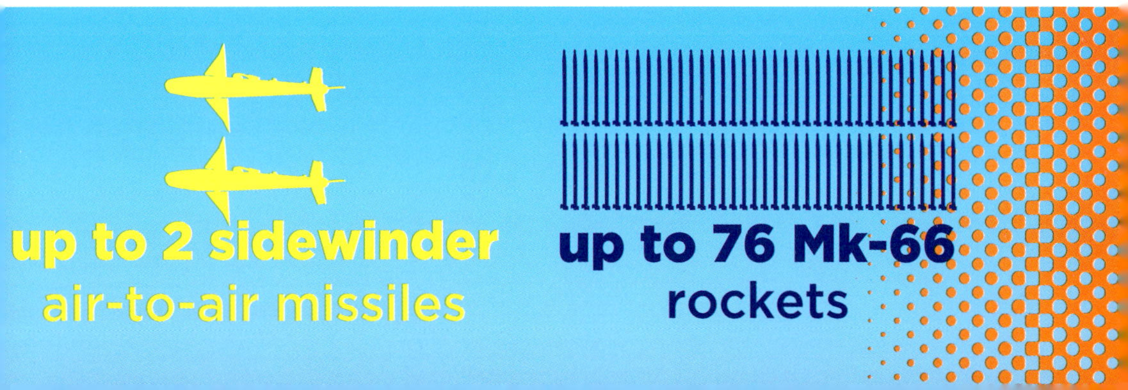

Cockpit

The Viper's glass cockpit has new technology. Pilots use a 3-D map to find their way. They use night vision to fly in the dark. Advanced **sensors** track faraway targets.

length
58.4 feet (17.8 meters)

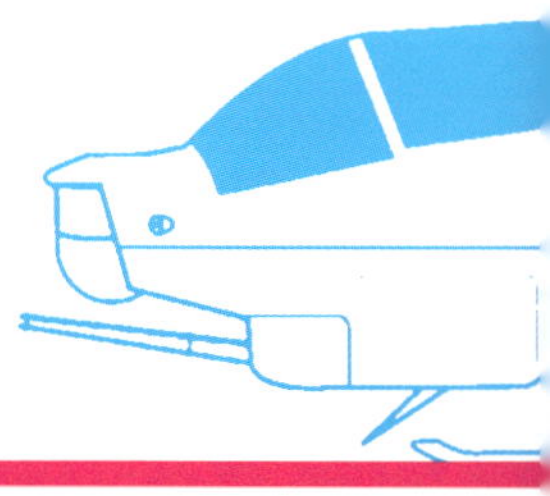

VIPER STATS

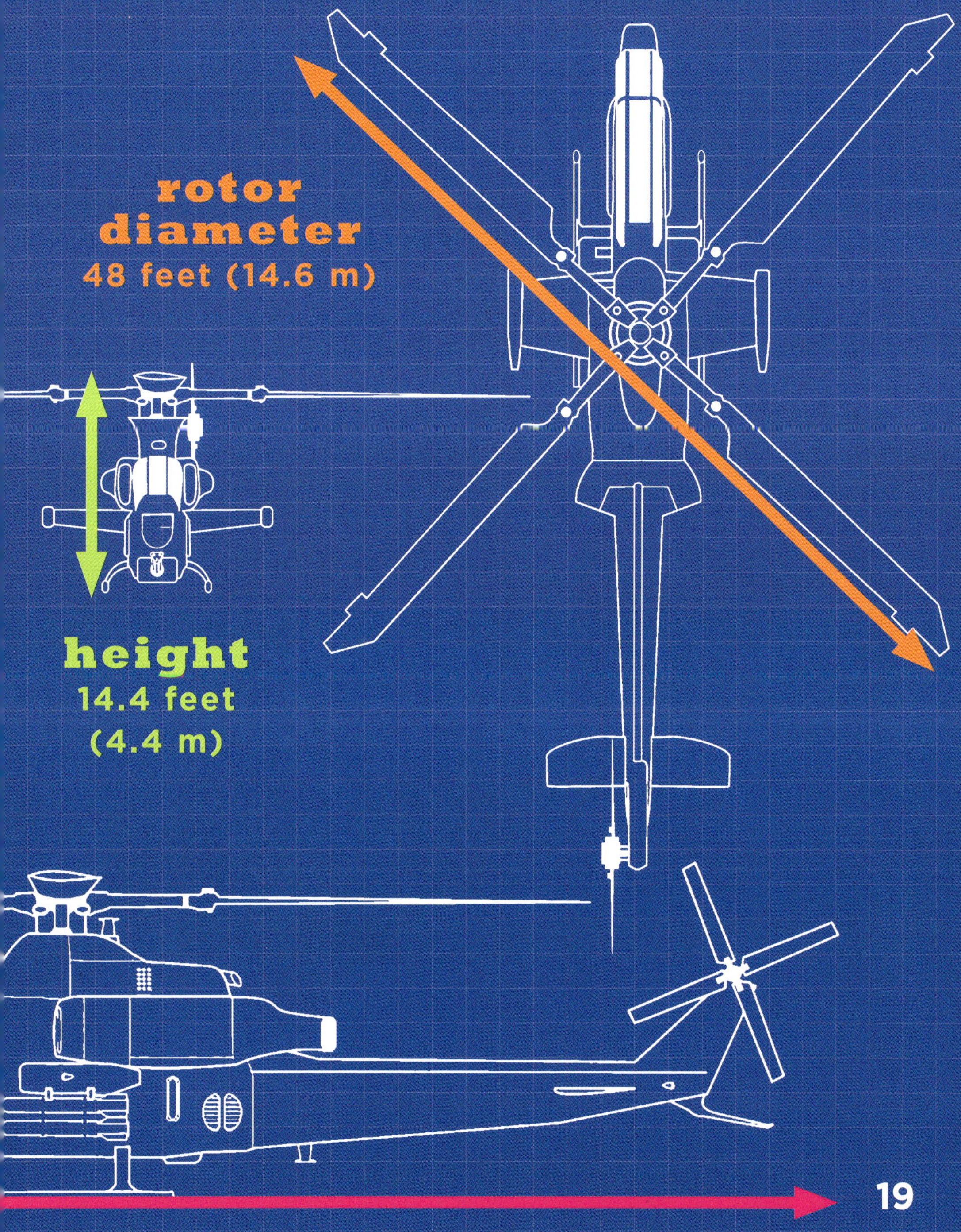

AH-1Z VIPER PARTS

MAIN ROTOR
COCKPIT
CANNON

CHAPTER 4

The Viper in ACTION

One of the Viper's jobs is to protect ground **troops**. It attacks enemies in the air and on the ground. The Viper's computers also send battlefield updates to soldiers. This information helps them decide where to go next.

Fighting Around the World

U.S. Marines use Vipers around the world. The countries of Pakistan and Bahrain are also buying the choppers.

The choppers might be used to fight **terrorists**. They will help in other missions too.

BY THE NUMBERS

20,000 feet
(6,096 m)
MAXIMUM FLYING HEIGHT

3.3 hours

MAXIMUM TIME VIPER CAN FLY BEFORE REFUELING

about
$30 MILLION
cost to buy
one Viper

85%

AMOUNT OF PARTS VIPER AND COBRA SHARE

Flying into the Future

The AH-1Z Viper helps Marines carry out missions. These choppers are fast and powerful. They will help keep soldiers safe for many years to come.

GLOSSARY

combat (kahm-BAT)—active fighting, often in a war

engineer (en-juh-NEER)—a designer or builder

hover (HUHV-er)—floating in the air without moving any direction

missile (MIS-uhl)—a weapon that can be thrown or projected to hit an object far away

rotor (RO-tuhr)—spinning blades that support a helicopter in the air

sensor (SEN-sor)—a device that finds heat, light, sound, motion, or other things

terrorist (TER-ur-ist)—a person who uses violent acts to frighten people in order to achieve a goal

troop (TROOP)—a group of soldiers

Vietnam War (vee-uht-NAHM WAHR)—a conflict between North Vietnam and South Vietnam that lasted between 1954 and 1975

LEARN MORE

BOOKS

Kiland, Taylor Baldwin, and Karen and Glen Bledsoe. *Military Helicopters: Heroes of the Sky.* Military Engineering in Action. New York: Enslow Publishing, 2016.

Peterson, Megan Cooley. *V-22 Osprey.* Air Power. Mankato, MN: Black Rabbit Books, 2019.

Whiting, Jim. *Night Stalkers.* U.S. Special Forces. Mankato, MN: Creative Education, 2018.

WEBSITES

AH-1Z Viper

www.navair.navy.mil/index.cfm?fuseaction=home.displayPlatform&key=B91F9D6E-E6AE-4F98-8BC5-A46CAE6C2381

Bell AH-1Z

www.bellflight.com/military/bell-ah-1z

The Bell AH-1Z Viper: Most Advanced Attack Helicopter in the World

www.youtube.com/watch?v=w98AMbjdthg

INDEX